Yolanda's

Cranberry Country Recipes

Best Wishes -
Yolanda Lodi

Yolanda's

Cranberry Country Recipes

by

Yolanda Lodi

Rock Village Publishing
Middleborough, Massachusetts

Typography and Title Illustrations
by Ruth Brown

ISBN 0-9674204-3-1

Dedication

To my best friend since childhood

my godmother

Rita Costa

Introduction

The people who live in Cranberry Country come from diverse backgrounds—English, Finnish, Polish, Irish, Asian, Italian, Cape Verdean, Portuguese, and any number of other ethnic groups. Over the years our cultures have intertwined, resulting in an eclectic taste in foods.

In *Yolanda's Cranberry Country Recipes* I try to capture the many flavors enjoyed by my family and friends, not only using cranberries grown on local cranberry bogs but also fruits and vegetables cultivated by local farmers—eggplants, kale, apples, rhubarb, blueberries, to name only a few. Because we live close to the ocean, seafood is readily available, either by catching your own (like Brad—see Cheryl's Seafood Lasagna) or purchasing it at a local market (such as Kyler's Catch Seafood Market—see Di's Baked Stuffed Shrimp).

My husband's love for cranberries (he grew up on the cranberry bogs), and other fruits, vegetables, and seafood inspired me to compile this cookbook to share with others. I hope you enjoy these recipes, and the little stories that embellish them.

Contents

Chicken • Seafood • and More

Chicken

Seafood

More

• Appetizers and Soups •

Appetizers

Soups

• Chutney and Sauces •

Chutney

Sauces

MUFFINS • BREADS • AND POPOVERS

Muffins

Breads

Popovers

• DESSERTS •

Miscellaneous

Crisps

Cookies

Pies

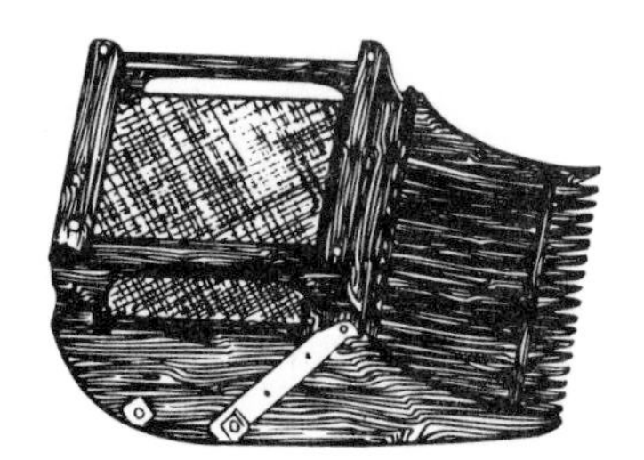

Chicken Seafood and More

CRANBERRY COUNTRY CHICKEN

1½ lbs. chicken breasts, boneless and skinless, cut in half and flattened
¼ cup bread crumbs

Sauce

1 cup fresh or frozen cranberries
½ cup sugar
⅓ cup onion, chopped
½ cup orange juice
⅛ tsp. ground cinnamon
⅛ tsp. ground ginger

Preheat oven to 400 degrees.

Stir *Sauce* ingredients in a saucepan over high heat. Bring to a full boil. (Cranberries will soften and split.) Remove from stove and set aside.

Coat chicken pieces with bread crumbs

Butter one oven-proof glass dish large enough to contain the four chicken breasts. Place coated chicken pieces in dish. Pour *Sauce* over chicken. Cover with aluminum foil.

Bake at 400 degrees for 30 minutes.

Serve over rice.

Makes 4 servings.

I often serve Cranberry Country Chicken when I've invited friends over for dinner. I prepare the sauce several hours beforehand, so that when my company arrives I spend less time in the kitchen and more time having fun with my guests.

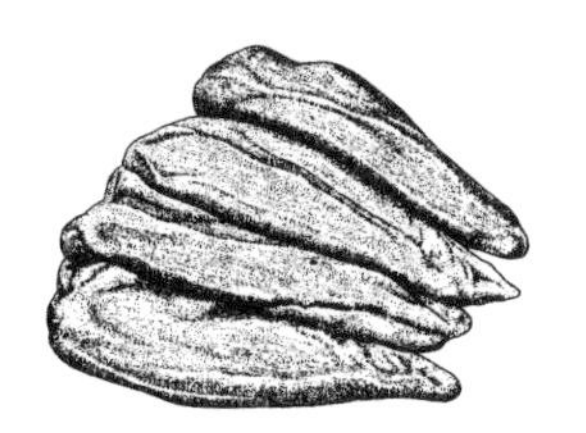

"Squeeze a Lemon" Chicken

1 ½ lbs.chicken breasts, boneless and skinless, cut in half and flattened
¼ cup bread crumbs
⅓ cup onion, chopped
½ cup chicken broth
1 tsp. dried parsley flakes
1 Tbsp. lemon juice (squeezed from half a lemon)

Preheat oven to 400 degrees.

Coat chicken pieces with bread crumbs.

Butter one oven-proof glass dish large enough to contain the four chicken breasts. Place coated chicken pieces in dish.

Combine onion, chicken broth, and parsley flakes in a bowl. Pour over chicken. Squeeze juice from half a lemon over the chicken pieces. Cover with aluminum foil.

Bake at 400 degrees for 30 minutes.

Serve over rice.

Makes 4 servings.

No lemon in the house? Although "Squeeze a Lemon" Chicken is more flavorful when made with freshly squeezed lemon juice, you can substitute juice from a bottle. Occasionally I do, when I find myself without lemons because of my drinking habits—my favorite beverage being cold water with a fresh squeeze of lemon.

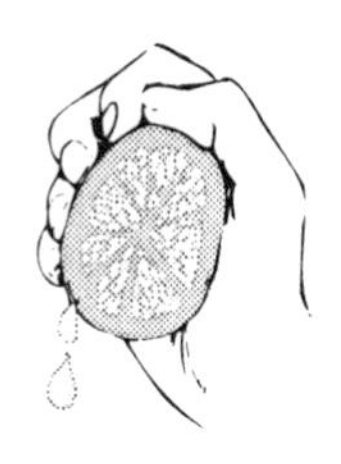

FATIMA'S CHICKEN MOZAMBIQUE

- 2 Tbsp. margarine
- 1 "handful" of chicken, cut in cubes (skinless and boneless)
- 1 clove garlic, chopped fine
- lemon juice squeezed from ¼ lemon
- pinch of red hot sauce (Red Hot)
- ½ cup white wine
- 1 packet Sazón Goya Con Azafran

Melt margarine in fry pan. Add all of the above ingredients *except* Sazón Goya Con Azafran. Bring to a boil.

While boiling stir in one packet Sazón Goya. Cook for about 5 minutes.

Serves 1.

Made to order.

The white wine and GOYA give the flavor. The spicy sauce comes from Red Hot.

Of all the many ethnic restaurants in Cranberry Country, Captain's Place in Acushnet, Massachusetts, is my favorite for Portuguese cuisine. Fatima's Chicken Mozambique is the dish for me—always cooked to perfection with plenty of spicy sauce for dunking fresh Portuguese bread. Oh, and the round fries, thinly sliced and golden brown . . . Delicious!

Fatima and her husband Joe serve great food in a family atmosphere. My husband and I are always looking forward to our next meal at Captain's Place and to chatting with the staff.

Make Fatima's Chicken Mozambique at home? Not I, when I can get it piping hot at Captain's Place.

Captain's Place, Fatima, and her Chicken Mozambique

Di's Baked Stuffed Shrimp

1½-2 lbs. jumbo tail-on shrimp, deveined (15 or under count per pound)

Stuffing

- 1½ cylinders Ritz crackers, crushed into cracker crumbs
- ¼ cup Parmesan cheese
- ½ cup (1 stick) butter, melted
- 1 Tbsp. lemon juice
- 1 Tbsp. brown sugar
- 1 tsp. cinnamon
- 2 Tbsp. plain yogurt
- 1 tsp. vanilla
- 3 Tbsp. water
- 1 large apple, finely chopped (Macintosh, or Granny Smith if you prefer a little more taste of tart)
- 1 cup finely chopped walnuts

Mix all *Stuffing* ingredients.

With tail upright, split the shrimp to "butterfly" them, forming a cavity for the stuffing. Arrange shrimp in a 9- x 13-inch pan (or appropriate size to contain all of the shrimp).

Stuff the shrimp, piling a generous portion of stuffing on each.

Bake at 350 degrees for approximately 20 to 22 minutes.

Serves 6

Ruth Brown's specialty is typography. Her talents in that field have helped make books from Rock Village Publishing a success. But when it comes to gourmet cooking, Ruth defers to her daughter-in-law, Diane Nanfelt.

Di's Baked Stuffed Shrimp is one of her favorites. Deveined jumbo tail-on shrimp are readily available—her husband owns Kyler's Catch Seafood Market in New Bedford, which specializes in fresh seafood, gourmet products, condiments, finishing sauces and pastas.

Next time you crave something special for dinner try Kyler's Catch Seafood Market and let your creative impulses run wild.

Di's Baked Stuffed Shrimp and Kyler's Catch Seafood Market

RITA'S CHRISTMAS COD

2 lbs. boneless salt cod fillets, cut into 4-inch pieces
½ cup olive oil
1 cup celery, chopped
1 cup onion, chopped
1 medium green pepper, chopped
½ tsp. crushed red pepper
¼ tsp. ground black pepper
1 can (8 oz.) tomato sauce

Two Days Before Cooking:

Place the salt cod in a large glass bowl and cover it with cold water. Cover the bowl tightly with plastic wrap. Soak the cod in the refrigerator for at least 2 nights, changing the water each morning and night.

Cooking Day:

Rinse the salt cod. Place the cod in a pan. Pour enough cold water to cover the cod.

Cover and cook over medium heat, bringing water to a boil.

Drain water.

Repeat this procedure again, bringing the water to a boil.

Using a fork, check cod. Cod is cooked when you can flake it with a fork.

Drain water and set aside to cool.

Sauce

In a skillet heat olive oil. Add all chopped vegetables, crushed red pepper, and black pepper.

When vegetables are *al dente* (partially cooked), add tomato sauce. Continue to cook until you can put a fork through the celery.

Cut the cooked cod into bite size pieces and place them in a glass serving dish.

Pour the *Sauce* over the cod. Sprinkle with fresh parsley.

Makes 4 to 6 generous servings.

My godmother, Rita, serves this cod every year on Christmas Eve. Last year I asked her for the recipe.

"I don't have a recipe. All you do is . . ."

I look forward to making Rita's Christmas Cod again. Now that I have her recipe I won't have to wait until Christmas Eve.

I must add that when I was a child,

Continued...

Christmas Eve at my godmother's was very special. Rita always had a beautifully decorated tree with lots of presents under it, including some for my brother and me. She knew that we wouldn't get many presents otherwise. I have fond memories of opening my presents while everyone had a great time, talking and laughing and enjoying the many foods spread out on the kitchen table. Cod was her specialty. And still is.

Months before Christmas friends ask Rita: "Are you making cod again this year?" The answer—to everyone's delight—is always, "Of course!"

HONEY BAKED SALMON

2 lbs. salmon fillets, cut into 4 serving pieces
2 Tbsp. honey
2 Tbsp. lime juice (juice from 1 large lime)
1 Tbsp. olive oil
1/8 tsp. ground black pepper

Coat a glass oven-proof baking dish with olive oil. Place the salmon fillets skin-side down.

Combine the honey, lime juice, olive oil, and pepper in a bowl. Brush the sauce over the salmon. Cover with aluminum foil and refrigerate for 30 minutes.

30 minutes later: Preheat oven to 450 degrees.

Remove salmon from refrigerator. Let stand at room temperature for 15 minutes.

Bake at 450 degrees for 15 to 20 minutes. Salmon is cooked when you can flake it easily with a fork.

Makes 4 servings.

Just before serving Honey Baked Salmon I remove the skin.

With a spatula I lift each salmon piece, then with a fork simply peel the skin off.

Scallops in a Pie Plate

1 lb. scallops
2 Tbsp. olive oil
¼ cup bread crumbs
2 Tbsp. sherry
1 tsp. minced garlic
ground black pepper (to taste)

Preheat oven to 450 degrees.

Coat a 9-inch glass pie plate with the olive oil.

Combine bread crumbs, sherry, and minced garlic in a small bowl. Roll scallops individually in the mixture until they are thoroughly coated.

Arrange scallops in pie plate.

Sprinkle ground black pepper to taste.

Bake at 450 degrees for 13 to 15 minutes.

Makes 2 servings.

A craving for scallops inspired me to make Scallops in a Pie Plate.

One afternoon on my drive from work I stopped and bought a pound of sea scallops. Home after a hectic day, I wanted something quick and easy and grabbed the first oven-proof dish that came to hand: a pie plate. "This will do nicely," I thought, and proceeded from there.

It took several subsequent experiments with Scallops in a Pie Plate to perfect this recipe to my taste, although if you ask my husband he'll probably say with a smile: "It was perfect the first time. Just be sure to double the portions."

Cheryl's Seafood Lasagna

Onion / Garlic Mixture

3 Tbsp. olive oil
5 garlic cloves, peeled and minced
1 cup white onion, chopped
2 Tbsp. dried basil
2 tsp. dried oregano
salt to taste
white pepper to taste

In a large skillet heat the olive oil over medium heat. Add garlic cloves and onions. Sauté until soft, but not browned (about 5 minutes). Stir in basil and oregano. Season with salt and pepper.

Remove from heat and set aside.

Béchamel Sauce

½ cup (1 stick) butter
½ cup all-purpose flour
4 cups milk
salt to taste
white pepper to taste

Melt butter in large saucepan over medium heat. Stir in flour. Cook, stirring constantly, for two minutes. Be careful *not* to brown.

Remove from heat and gradually stir in milk. Blend until smooth.

Return to heat and stir constantly until thickened. Season with salt and pepper.

Remove from heat.

Makes 4 cups.

Set aside 1 cup of Béchamel Sauce for later.

Seafood Sauce

- 3 cups Béchamel Sauce
- ½ cup Parmesan cheese, freshly grated
- 3 Tbsp. dry sherry
- 1 lb. medium shrimp, cooked, peeled, and deveined
- 1 lb. bay scallops, cooked
- salt to taste
- white pepper to taste

Add the *Onion / Garlic Mixture* to the remaining 3 cups of Béchamel Sauce.

Stir in the Parmesan cheese and dry sherry. Heat gently.

Fold in shrimp and scallops. Season with salt and pepper. Set aside.

Continued...

CHERYL'S SEAFOOD LASAGNA...*Continued*

Sautéed Mushrooms

1 lb. fresh mushrooms, sliced
2 Tbsp. butter

In a large skillet sauté the mushrooms in hot butter over medium heat, until the mushrooms are tender and the liquid has evaporated.

Set aside.

Seafood Lasagna

1 lb. lasagna noodles, cooked according to package instructions
1 pkg. frozen chopped spinach (10 oz.), thawed and well drained
1 lb. lobster meat, cooked and cut into bite-size chunks
1 lb. mozzarella cheese, freshly grated

Grease the bottom of a 9- x 13-inch glass oven-proof baking dish.

Spread a very thin layer of *Béchamel Sauce* (from the reserved 1 cup) on the bottom. Cover with a layer of cooked lasagna noodles.

Spread half of the *Seafood Sauce* over the noodles. Top with half of the chopped spinach, one-fourth of the mozzarella cheese, half of the *Sautéed Mushrooms*, and half of the lobster.

Repeat the layers, beginning and ending with the lasagna noodles.

Spread the remaining *Béchamel Sauce* over the noodles.

Top with the remaining mozzarella cheese.

Bake *uncovered* at 350 degrees for 45 to 60 minutes, until bubbly and lightly browned.

Let stand for 15 minutes before serving.

Makes 10 to 12 servings.

The town of Middleborough celebrates the Christmas holiday with a smashing parade, Santa Claus, hot cider, and all the warm inviting pleasures of the season.

The Middleboro Gazette *announced that the Zachariah Eddy House Bed and Breakfast located next to the Town Hall was open for tours and refreshments. Ed and I not only love old Victorian houses, but we also enjoy staying in them while vacationing.*

We approached the massive doors where Cheryl—whom we'd never met before—greeted us like old friends, smiling warmly ear-to-ear as she shook our hands,

Continued...

saying: "Be sure to visit the Chapel upstairs. We've converted it to a much needed bathroom for guests. And come down and enjoy some homemade goodies by the fire."

We followed the crowd, simultaneously blurting out: "What a cheerful person. And what a great B&B!"

At the end of our tour Ed and I chatted with Cheryl about the history of the home and how much we enjoyed her hospitality. Before we left Ed said: "We love B&B's."

Cheryl placed her hands on Ed's bearded face and gave him a quick kiss, saying: "I just love you."

The following year Ed and I and our twenty-two guests celebrated our marriage in Cheryl's beautiful B&B.

The main entrée at our wedding was Cheryl's Seafood Lasagna, loaded with lobster from her husband Brad's lobster traps. To quote Cheryl: "This spectacular dish is a labor of love reserved for special occasions. Costly ingredients and lengthy preparation time take it out of the realm of the everyday. Even so, it's well worth the effort for something truly special."

Cheryl's Seafood Lasagna is memorable gourmet eating at its best.

Zachariah Eddy House Bed and Breakfast
Middleborough, Massachusetts

BAKED EGGPLANT FOR TWO

- 1 large eggplant
- 1 Tbsp. bread crumbs
- ½ Tbsp. Italian seasoning
- 1 tsp. minced garlic
- ⅛ tsp. ground black pepper
- 1 large ripe tomato, peeled and thinly sliced
- 2 Tbsp. 100% pure olive oil

Wash eggplant. Cut in half lengthwise. Sprinkle with salt. Drain in a colander for one to two hours to rid eggplant of bitter flavor.

Preheat oven to 375 degrees.

When ready to bake, wipe eggplant dry with paper towel. Make short deep cuts on each half with the point of a sharp knife.

Mix bread crumbs, Italian seasoning, minced garlic, and pepper. Spread mixture on eggplant, filling the cuts.

Arrange eggplant cut-side-up in glass baking dish. Cover with tomato slices. Pour 1Tbsp. olive oil over each half. Cover with aluminum foil.

Bake at 375 degrees for 40 minutes or until eggplant is tender.

Serves two.

One year my husband's garden produced an over abundance of eggplants. Dozens of these purple-skinned vegetables hung from the plants, ready for picking.

How was I going to prepare these? After some trial and error I came up with what my husband and I both agree is a healthful, tasty treat.

Since that year the trees around our garden have grown taller, creating too much shade to grow vegetables. Because we choose not to cut down the trees, we have converted the area to a flower garden, where the likes of jacks-in-the-pulpit, bleeding hearts, and pulmonaria flower and flourish.

Because the flowers are so beautiful, I remember our former vegetable garden, and how I came to create this recipe, only when I prepare Baked Eggplant for Two.

INDIAN PIZZA

Crust

1¼ cups all-purpose flour
¾ cup yellow cornmeal
1 tsp. baking powder
⅔ cup skim milk
¼ cup olive oil

Preheat oven to 425 degrees.

Coat a 14-inch round pizza pan with non-stick cooking spray.

In a bowl, combine flour, cornmeal, and baking powder and mix well. Add milk and olive oil. Stir with a spoon to form a ball.

Place ball in center of pizza pan. Wait 5 minutes (just enough time to wash the items used to prepare this crust).

Using the palm of your hand spread dough evenly toward the edge of the pan, forming a rim at the edge. Bake at 425 degrees for 13 minutes or until golden brown.

Remove from oven. Now you're ready to create!

Suggestions:

Spread your favorite pizza sauce (or your favorite *thick* spaghetti sauce) over the crust. Top with 2 cups of your favorite shredded cheese(s). On top of the cheese place cut-up vegetables, such as onion, pepper, mushrooms, broccoli.

WORD OF WARNING: Do not top with meat. Crust may become soggy.

Return pizza pan to oven. Bake at 425 degrees 15 to 20 minutes, until cheese is melted and vegetables are cooked to your taste.

Cut into 8 pieces and enjoy a healthful treat.

Craving a pizza tonight?

With this recipe you don't have to wait for the crust to "proof." The proof is in the eating.

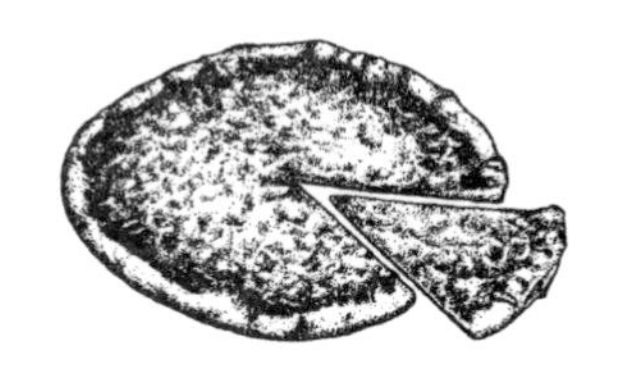

Passionate Pesto

- ½ cup dried basil leaves
- ½ cup grated Parmesan cheese
- ½ cup walnuts
- ½ tsp. garlic powder
- ¼ tsp. salt
- ½ cup 100% pure olive oil
- 1 Tbsp. lemon juice

Prepare food processor for chopping.

Place all ingredients *except* olive oil and lemon juice in work bowl. Process until combined (about 15 seconds).

With food processor running, pour olive oil through feed tube. Process until combined. Scrape sides of working bowl. Continue to process for another 15 seconds, adding lemon juice through feed tube.

Refrigerate until ready to use with your favorite cooked pasta.

This recipe makes enough pesto
for ¾ lb. (uncooked) pasta.

If you love pesto as much as I love pesto, you'll be passionate about this recipe.

After trying store-bought pesto, woefully lacking in basil and nut flavor, I created Passionate Pesto, in the belief that "Eating well is the best revenge."

Speaking of eating well, La Famiglia Giorgio's, a great restaurant in Boston's North End, serves a superb pesto which my Italian husband—ever the diplomat—says is the best he's ever tasted, other than mine, of course!

Appetizers and Soups

LANGOSTINO ON THE HALF SHELL

1 package (12 oz.) Langostino
2 cylinders Ritz crackers, broken into pieces
1 medium onion, chopped
2 cans (6.5 oz. each) chopped clams with juice

½ cup (1 stick) butter or margarine, melted
¼ cup skim milk
½ tsp. garlic powder
¼ tsp. dried parsley flakes

Note: For this recipe you will need 12 quahog or scallop shells, or similar size dishes.

Preheat oven to 350 degrees.

Prepare food processor for chopping.

Place the first four ingredients (langostino, crackers, onion, clams) in *work bowl*. Pulse until combined.

In a large bowl whisk the remaining four ingredients until creamy. Add *work bowl* mixture. Stir until combined thoroughly.

Fill 12 shells. Place filled shells on a cookie sheet. Bake at 350 degrees for 30 minutes.

Allow shells to cool to the touch before serving.

I prefer scallop shells for this recipe. Where did I obtain them?

My husband and I bought stuffed scallops at the local market. The shells were gorgeous, so we kept them. What a treat, twice around! Or you may think that finding your own at the beach is more fun.

KILLER QUICHE

4 Tbsp. (½ stick) butter or margarine
3 large eggs
1 cup all-purpose flour
1 cup skim milk
½ tsp. salt
1 tsp. baking powder
1 lb. grated white sharp cheddar cheese
2 pkg. frozen chopped spinach (10 oz. each), thawed and well drained *(I squeeze the spinach dry with my hands.)*
½ cup onion, chopped

Preheat oven to 350 degrees. Melt butter in 9- x 13-inch glass oven-proof baking dish. Tilt to coat sides.

In a large bowl beat eggs until foamy. Add flour, skim milk, salt, and baking powder. Mix until all ingredients are combined.

Add cheese, spinach, and onion. Mix well with a spoon. Spread into buttered dish.

Bake at 350 degrees for 35 minutes.

When cooled cut into bite-size squares.

Leftovers, if any? Can be frozen for future treat.

I bet you can't eat just one! I'm not a lover of spinach but I love my Killer Quiche. As my husband likes to say, "Give me a little quiche, huh, will yuh?"

DEBBIE'S CLAM CHOWDER

4 cans (6.5 oz. each) minced clams
1 can (10 oz.) whole baby clams
1 large onion, chopped
4 cups potatoes, peeled and diced
3 Tbsp. flour
1 cup milk
1 container (16 oz.) light cream
½ cup (1 stick) butter
Seasonings to taste:
salt
pepper
dill weed (use generously)
dill seed (use generously)
basil (a pinch)
thyme (a pinch)

Drain clam juice from all 5 cans into a large pot. Add onion and *1 cup of water*. Bring to a boil.

Add potatoes. Cook 15 to 20 minutes over medium heat.

Add clams, flour, milk, cream, butter, and seasonings. Reduce heat to simmer. (Do not boil.) Cook slowly until potatoes are tender but not mushy.

Debbie's Suggestions

- Best when made a day ahead. Simply refrigerate when done and reheat on low. (Do not boil.)

- For a thicker chowder, add more flour.
- For a heartier chowder, add chunks of cooked lobster, fish, scallops, or shrimp. (Do not overcook.)
- For a lower fat content, substitute margarine for butter and use 1% milk instead of whole.
- Serve with oyster crackers or fresh bread.

Makes 10 to 12 servings.

As busy as Debbie is, helping her husband Bob run the Duff Gallery in New Bedford, she always manages to find time to make her popular Clam Chowder for family and friends.

In Debbie's own words: "This is a no-fail recipe. There's a subtle difference to the flavor every time, depending on the amount of seasonings I sprinkle into the pot. I use a lot of dill weed and dill seed to give the chowder color and a rich flavor."

A "no-fail" recipe it is. The first time I made Debbie's Clam Chowder I grabbed the basil—thinking it was one of the dills—and added a teaspoon instead of a pinch. When I realized my mistake it was too late. So I decided to add a teaspoon each of the dills, along with one-fourth of a teaspoon of ground black pepper, and to omit the thyme and the salt. The chowder came out perfect!

CHEESEY BROCCOLI SOUP

1 medium onion, chopped
¼ tsp. salt
¼ tsp. ground black pepper
¼ tsp. dried thyme flakes
⅛ tsp. ground nutmeg
2 cups skim milk
1 can (14.5 oz.) chicken broth
1½ lbs. broccoli, sliced thin
1½ cups grated white sharp cheddar cheese

Microwave onion and *1 tsp. of water* for 2 minutes.

Add salt, pepper, thyme, and nutmeg. Stir to coat onion. Microwave an additional 1 minute.

Coat a large soup pan with non-stick cooking spray.

Place onion mixture, milk, chicken broth, and *1½ cups water* in pan. Stir occasionally over medium heat. Before mixture begins to boil, add broccoli, then increase heat to high.

Heat to boiling, stirring frequently. Reduce heat to low. Cover and simmer for 10 minutes until broccoli is tender. Cool.

In a food processor, purée cooked broccoli and onions, gradually adding liquid from soup pan through feed tube.

Return purée to soup pan. Heat soup to boiling over high heat, stirring frequently.

Remove pan from heat. Stir in cheese until melted and smooth.

Makes 8 servings.

Say "cheese" and smile. Broccoli, my favorite vegetable, appears on our shopping list weekly. And when broccoli is on sale, you can bet I cook up a pan of Cheesey Broccoli Soup.

I hope it puts a smile on your face, too.

SPICY CARROT SOUP

1 medium onion, chopped
2 tsp. curry powder
½ Tbsp. ground ginger
1½ lbs. carrots, sliced thin
1 can (14.5 oz.) chicken broth
½ tsp. salt
½ cup half-and-half

Microwave onion and *1 tsp. of water* for 2 minutes.

Add curry powder and ground ginger to cover onions. Microwave an additional 1 minute.

Coat a large soup pan with non-stick cooking spray.

Place onion mixture, carrots, chicken broth, salt, and *1 cup water* in pan. Heat to boiling, stirring frequently. Reduce heat to low. Cover and simmer for 20 minutes until carrots are tender. Cool.

In a food processor purée cooked carrots and onions, gradually adding liquid from soup pan through feed tube.

Return purée to soup pan. Stir in half-and-half and *2 cups water*.

Heat soup over medium heat until hot, stirring frequently.

Makes 8 servings.

On a frigid winter day hot spicy soup hits the spot. The aroma of the spices warms the soul as well as the heart.

CALDO VERDE

- 1 medium onion, chopped
- 1 tsp. minced garlic
- 1 lb. all-purpose potatoes, peeled and cut into 2-inch pieces
- 1 can (14.5 oz.) chicken broth
- ¼ tsp. salt
- ⅛ tsp. ground black pepper
- ½ lb. kale, sliced thin (remove coarse stems and veins)

Microwave onion and *1 tsp. of water* for 2 minutes.

Add minced garlic. Stir to coat onion. Microwave an additional 1 minute.

Coat a large soup pan with non-stick cooking spray.

Place onion mixture, potatoes, broth, salt, pepper, and *1½ cups water* in pan. Heat to boiling. Reduce heat to low. Cover and simmer for 20 minutes until potatoes are tender. Remove from heat.

With a potato masher, mash potatoes until they are lumpy. Stir in kale.

Return to stove. Simmer, uncovered, for about 5 minutes, until kale is tender.

Makes 4 servings.

As a child I turned up my nose whenever my mother served Caldo Verde. Maybe it was all that green, *reminding me of spinach. Who knows?*

As an adult I enjoy a hot bowl of Caldo Verde with fresh Portuguese bread—so much so, I can make a meal of it. How our taste buds change for the better!

Next time fresh kales are available give Caldo Verde a try.

Chutney and Sauces

SOPHIE J.'S CRANBERRY CHUTNEY

4 cups fresh cranberries
½ cup golden raisins
½ cup cider vinegar
½ cup brown sugar
½ cup green pepper, chopped
½ cup mixed candied fruit and peels (fruit cake mix)
¼ tsp. ground clove
¼ tsp. ground ginger
¼ tsp. ground cinnamon
¼ tsp. ground allspice

In a medium size pan combine all ingredients. Bring to a boil. Reduce heat and simmer for 30 minutes, stirring!

Cool.

Refrigerate in airtight containers. Will keep for 6 to 8 weeks.

Cranberry Country has many fascinating places to visit, among them the Old Company Store in Wareham, Massachusetts, a historic building that was once part of a nail factory and is now a unique gift shop full of treasured specialty items and gourmet foods, along with make-you-feel-at-home ladies Shari and Cindi (owners and sisters), Sandi (their mother), and Sophie, to name a few.

Recently, while I was sampling one of their flavored coffees, Sophie mentioned her Cranberry Chutney and how everyone enjoys it. The next thing I knew, not only was I bringing home two pounds of ground coffee, but the recipe for Sophie J.'s Cranberry Chutney as well.

Cherry Berries

4 cups firm Cranberries
2 cups sugar
1 cup water
1/4 Tsp baking soda
1/4 Tsp salt

Mix together and bring to a boil in large pot. Cover pot tightly and simmer 15 min.

Do not remove lid at any time.

Remove lid ONLY when pot becomes STONE COLD.

Great as a sauce with Turkey, chicken, Pork etc. Also Tasty over ice Cream, Pound cake, or "what have you"

Dorthy "Dude" Massey gave me the recipe for Cherry Berries just as you see it, in her own distinctive calligraphy.

Be sure to follow her instructions to the letter. Resist temptation! Do not transfer the sauce from the pan to clear glass jars until the next day.

Cherry Berries look and taste like cherries . . . cheerily red and succulently sweet. As Dorthy says: tasty over ice cream, pound cake, or what have you.

RUBY PORT SAUCE

3 cups cranberries
1 cup Port wine
¾ cup sugar
½ tsp. ground cinnamon
¼ tsp. ground nutmeg
⅛ tsp. ground allspice

Place all ingredients in a 2-quart saucepan. On high heat stir to boiling.

Reduce heat to medium. Continue to stir and cook until most cranberries pop.

Remove from stove. Cool. Spoon sauce into airtight container.

Refrigerate for at least 3 hours so that mixture thickens.

Makes about 2 cups.

Serving chicken or turkey for the holidays? Add this cranberry sauce to your festive dinner. Your friends will sail into this port any day!

APPLE PORT SAUCE

- 7 medium cooking apples, peeled and coarsely chopped
- ½ cup Port wine
- ½ tsp. ground cinnamon
- ⅛ tsp. ground nutmeg

Place all ingredients in a 2-quart saucepan.

On medium heat stir to boiling. Reduce heat to simmer. Stirring occasionally, cook for 5 to 10 minutes or until apples are tender.

Remove from heat. Mash with a potato masher. Stir and pour into a glass dish. Ready to serve or refrigerate.

Makes about 2 cups.

My Ruby Port Sauce gave me the idea for this recipe. I love applesauce, but not out of a jar, where it's either watery or too sweet.

Apple Port Sauce smells wonderful while cooking. It tastes perfectly sweet with no added sugar. And it's thick enough so that you can serve it as an added treat with any festive dinner.

It's also great as a healthful snack.

Muffins Breads and Popovers

Valentine's Day Muffins, Every Day

- ½ cup (1 stick) butter or margarine, softened
- 1 cup sugar
- 2 large eggs
- 2 cups all-purpose flour
- 2 tsp. baking powder
- ¼ tsp. salt
- ½ cup skim milk
- 1 tsp. vanilla extract
- 2½ cups cranberries, fresh or frozen (thawed), cut in half
- 2 Tbsp. chopped walnuts

Preheat oven to 350 degrees.

Grease twelve 2½-inch-muffin-cup pan well with butter, including the top surface of pan.

In a large bowl, with mixer at low speed, cream butter and sugar until smooth and fluffy. Add one egg at a time, making sure batter continues to be fluffy.

Increase speed to medium. Add flour, baking powder, and salt, alternately with milk, until blended. Stir in vanilla.

Add cranberries. Stir with a spoon until combined.

Spoon batter into muffin cups. Sprinkle chopped walnuts on top.

Bake at 350 degrees for 40 minutes.

Makes 12 muffins.

Cranberries are tart
Strawberries are sweet
These Valentine Muffins
can't be beat!

Cranberry Nut Bread

- 3 cups all-purpose flour
- 1 cup sugar
- 4 tsp. baking powder
- ½ tsp. salt
- 1 large egg
- 1½ cups skim milk
- 2 Tbsp. olive oil
- 1¼ cups cranberries, fresh or frozen (thawed), cut in half
- ¾ cup chopped walnuts

Preheat oven to 350 degrees.

In a large bowl combine flour, sugar, baking powder, and salt. Mix with a spoon.

In a separate medium-size bowl whisk egg. Then add skim milk and olive oil. Whisk until foamy.

Add to dry ingredients. Mix to form a batter. Fold in cranberries and walnuts. Spread into greased and floured 9- x 5- x 3-inch loaf pan.

Bake at 350 degrees for 1 hour or until toothpick inserted in center comes out clean. Cool 15 minutes then remove from pan.

Makes multiple servings
(exactly how many depends on how you slice it).

This is my favorite bread. It's easy to make, loaded with cranberries and walnuts, and always comes out perfect.

There's another reason why Cranberry Nut Bread is a favorite of mine.

It evokes childhood memories of Edaville Railroad at Christmas time—the snow-covered grounds, Yuletide music, train whistles, and magical rides…the train chugging along past festive scenes, little houses decorated with colored lights, storybook people and animals dressed for the occasion, a lighthouse with its shining beacon, trees and shrubs decked out in holiday finery…a surprise at every bend in the tracks.

Then, after the ride, purchasing a loaf of cranberry bread to take home. This was a special treat for all. To my parents, having immigrated from Pico, Azores, cranberries had always meant only canned sauce at Thanksgiving. To think—there was also a delicious bread made with cranberries!

Last year when Edaville Railroad re-opened I rode the train again. As a child I had never guessed that there were cranberry bogs under the snow. But this time it was harvest season. What a spectacular show of color—red shining berries floating in the water against a backdrop of brilliant autumn foliage at its peak. Seeing this

Continued…

natural beauty, the bogs and the wild creatures that make the bogs and the wetlands their home, makes me realize, once again, how wonderful and exciting it is to live in Cranberry Country.

Whew! All this from a loaf of Cranberry Nut Bread!

BREW BREAD

3 cups self-rising flour
1 cup ale (or beer)
½ cup sugar
4 Tbsp. (½ stick) butter or margarine, melted

Preheat oven to 350 degrees.

Coat a 9- x 5- x 3-inch loaf pan with non-stick cooking spray.

Mix flour, ale, and sugar with a spoon to form a large ball. Spread in loaf pan. Bake at 350 degrees for 45 minutes.

Poke holes in top crust with the point of a sharp knife. Pour melted butter over top of bread. Return pan to oven. Bake an additional 15 minutes.

Cool 15 minutes then remove from pan.

Makes 1 loaf.

Brew Bread, my husband's favorite—the old buzzard. No kidding. I use Olde Buzzard ale from Buzzards Bay Brewing of Westport, Massachusetts, to make this bread.

Maple Brown Bread

1 cup all-purpose flour
1 cup whole-wheat flour
¾ cup raisins
¼ cup sugar
1¼ tsp. baking soda
¼ tsp. salt
1 Tbsp. lemon juice *plus*
skim milk *to total 1¼ cups liquid mixture of lemon juice and skim milk* (Stir. Let mixture stand for 5 minutes.)
¾ cup maple syrup
1 large egg

Preheat oven to 350 degrees.

In a *large bowl* combine flours, raisins, sugar, baking soda, and salt. Mix with a spoon.

In another bowl stir liquid mixture (lemon juice / skim milk), maple syrup, and egg. Add to *large bowl* and stir until mixed. (Batter will be very soupy.)

Pour batter into greased 9- x 5- x 3-inch loaf pan.

Bake at 350 degrees for 1 hour until toothpick inserted in center comes out clean. Cool 15 minutes then remove from pan.

Makes 1 loaf.

A gift of maple syrup suggested this recipe. My husband and I received a jug of maple syrup from a dear friend. What do you do with a jug of maple syrup? It's good for you but can you cook with it? Of course: Maple Brown Bread!

QUICK AND EASY POPOVERS

1¼ cups all-purpose flour
1 cup skim milk
3 large eggs
2 Tbsp. (¼ stick) butter or margarine, melted
¼ tsp. salt
¼ tsp. ground black pepper
⅓ cup dried onion flakes

Preheat oven to 400 degrees. Spray twelve 2 ½-inch-muffin-cup pan with non-stick cooking spray.

Place all ingredients except dried onion flakes in a large bowl. Mix at medium speed for about 2 minutes, scraping sides. Batter will be soupy.

Stir in onion flakes. Spoon or pour evenly into muffin cups.

Bake at 400 degrees for 30 minutes or until puffed and well browned.

Remove from oven. Tilt muffin pan until popovers fall onto wire rack. Using the tip of a small knife pierce each popover to release steam.

Best served hot.

Makes 12 popovers.

Popovers require T.L.C. to ensure that they pop.

On one occasion I forgot the melted butter, *and added it at the end of the recipe. My popovers came out looking like deflated tires.*

Another time I waited too long to beat the ingredients, so that the butter thickened and the batter was not sufficiently soupy. The popovers came out looking like perfect dinner rolls—delicious, but not eye-bulging treats.

Desserts

CRANBERRY COUNTRY CHEESECAKE

Crust

1¼ cups graham cracker crumbs (11 rectangular graham crackers)
4 Tbsp. (½ stick) butter or margarine, melted
1 Tbsp. sugar

Filling

3 packages (8 oz. each) cream cheese, softened
¾ cup sugar
1 Tbsp. all-purpose flour
1½ tsp. vanilla extract
3 large eggs
1 egg yolk
¼ cup skim milk

Crust

Preheat oven to 375 degrees.

Prepare food processor for chopping.

Place *Crust* ingredients in *work bowl*. Pulse until mixture is crumbs. Using your fingers, press the buttered crumbs evenly and firmly onto the bottom and up the sides of an 8½-inch springform pan. Bake at 375 degrees for 10 minutes.

Cool on rack.

Lower oven temperature to 300 degrees.

Filling

In a large bowl, with mixer at medium speed, beat cream cheese and sugar until mixture is smooth and fluffy. Beat in flour and vanilla until combined.

Reduce speed to low. Add 1 egg at a time, including egg yolk, beating after each addition. Beat in milk until blended.

Pour batter into springform pan.

Bake at 300 degrees for 55 to 60 minutes until lightly golden and the area 3 inches from the center is slightly wet.

Cool completely on wire rack.

Refrigerate overnight.

Cranberry Topping

Note: Cranberry Topping can also be used as a fruit spread on crackers and toast, or as a cranberry sauce with chicken and turkey.

- 2 cups cranberries, fresh or frozen, cut in half
- ⅔ cup sugar
- 1 Tbsp. cornstarch
- ¼ cup orange juice with pulp

Place all ingredients in a 2-quart saucepan. *Continued...*

CRANBERRY COUNTRY CHEESECAKE...*Continued*

On high heat stir to boiling. Reduce heat to medium. Continue to stir and cook until most cranberries pop and mixture has the consistency of jam (*about 3 minutes*).

Remove from heat. Cool completely. Refrigerate overnight.

The next day

Place cheesecake on a plate. Remove side of springform pan. Garnish with *Cranberry Topping*.

Makes 12 servings.

Before removing the side of the springform pan, I place my Cranberry Country Cheesecake on a 12-inch decorative glass plate with turned-up edges. I do this for two reasons. The edges protect the cheesecake from slipping off the plate (OOPS!). And a fancy dish completes the festive, ice-cream-sundae look of the rich, red cranberries against the creamy-vanilla color of the cheesecake.

Every bite is delightful: so rich and smooth, with just the right degree of sweetness, that I forget the topping is made with cranberries, which at times can be tart. But not in this recipe, where every bite has a pucker-up, refreshing taste.

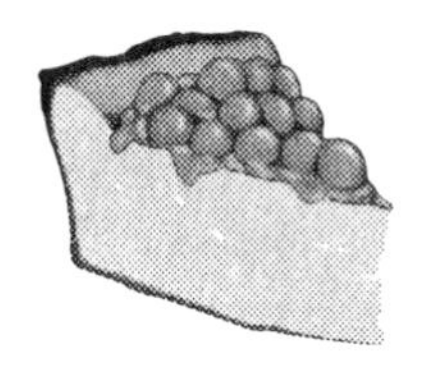

CRANBERRY TOPSY-TURVY

Topping

4 Tbsp. (½ stick) butter or margarine
¾ cup packed dark-brown sugar
1½ cups fresh or frozen cranberries (thawed)
½ cup golden raisins, plumped

Cake

½ cup (1stick) butter or margarine, softened
¾ cup packed dark-brown sugar
1 Tbsp. vanilla extract
1½ tsp. baking powder
2 large eggs
⅔ cup skim milk
1¼ cups all-purpose flour
⅓ cup yellow cornmeal

Topping

Preheat oven to 350 degrees.

Place butter in 8- x 2-inch round glass oven-proof baking dish. Melt butter in oven. Remove from oven. Tilt to coat sides.

Sprinkle brown sugar evenly over melted butter. Then place cranberries evenly in one layer. Finish with raisins.

Cake

In a large bowl, with electric mixer on medium speed, beat 3 minutes butter, brown sugar, vanilla, and baking powder. Mixture will be very pale. Continue to beat adding 1 egg at a time. Mixture will look curdled.

On low speed beat in milk, flour and cornmeal until blended.

Spread evenly over *Topping*.

Bake at 350 degrees for 55 to 60 minutes until top is browned. Test by inserting toothpick in center. Ready when pick comes out with moist crumbs.

Cool in pan on wire rack for 5 minutes before inverting cake onto serving plate.

Makes 12 servings.

If you plan to use frozen cranberries in this recipe be sure they are completely defrosted first. Don't make the mistake I made once. I used cranberries that were still frozen, and my "topsy-turvy" turned out to be a "flip-flop," edible but not memorable. The cake did not rise.

Beth's Bread Pudding

1 loaf (10 oz.) stale French bread, broken into bite-size pieces

or

combination stale French bread and stale wheat bagel or stale wheat bread to total 12 cups bite-size pieces

2 cups milk
2 cups heavy cream
2 cups sugar
½ cup (1 stick) butter, melted
3 large eggs
2 Tbsp. vanilla extract
2 cups shredded coconut
2 cups chopped pecans or walnuts
1 can (8 oz.) crushed pineapple, drained
2 tsp. ground cinnamon

Preheat oven to 350 degrees.

Butter a 9-x-13-inch glass oven-proof baking dish.

Combine all ingredients in a large bowl. Mixture will be very moist. If soupy, add more bread pieces.

Let stand for 10 to 15 minutes, until bread is well soaked.

Pour into buttered baking dish. Place on middle rack of oven. Bake at 350 degrees for approximately one hour or until top is golden brown.

Serve warm with sauce. (*See following page for Lemon Sauce recipe.*)

Makes 10-12 servings.

Beth's Bread Pudding is an extra special treat—rich, sweet, loaded with coconut and nuts, sometimes pecans, other times walnuts. (My preference is pecans, with a little pineapple added. Try it either way. You be the judge!) This dessert brings back fond memories of our travels with Beth to New Orleans, where Bread Pudding is a specialty.

Whenever I make this dessert for my slightly *overweight husband and me, I make the following changes to Beth's recipe:*

4 cups skim milk *instead of* 2 cups milk and 2 cups heavy cream

1½ cups sugar *instead of* 2 cups sugar

½ cup margarine (1 stick) *instead of* ½ cup butter

I look forward to Beth's Bread Pudding whenever my husband and I get together with her. It's always a taste of New Orleans when dinner is topped off with her fabulous Bread Pudding. As Beth likes to say, with that sweet smile of hers: "I left out the fat and calories."

Lemon Sauce

for Beth's Bread Pudding

½ cup (1 stick) butter
1½ cups powdered sugar
1 egg
½ cup lemon juice

Cream butter and sugar over medium heat until smooth and creamy.

Remove from heat. Blend in egg. Continue to stir while adding lemon juice.

Serve warm over Beth's (warm) Bread Pudding.

Note: Sauce thickens when cooled. Warm sauce before serving.

Best when made the day before.

Beth's Bread Pudding and Lemon Sauce freeze well.

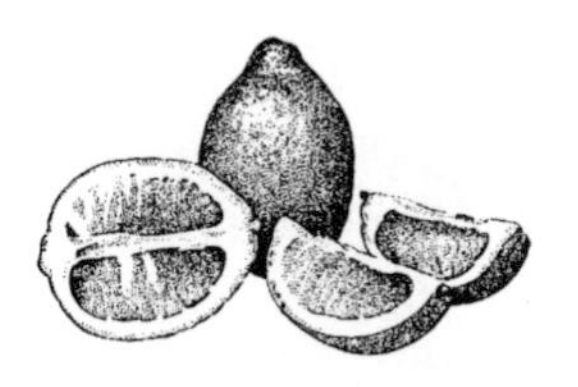

GRANNY'S CRANBERRY CRISP

Topping

6 Tbsp. (¾ stick) butter or margarine, softened
¾ cup rolled oats
¾ cup all-purpose flour
½ cup sugar

Combine all *Topping* ingredients in a bowl. Mixture will be crumbly. Set aside.

Filling

5 large Granny Smith apples, peeled and coarsely chopped
1 cup cranberries, fresh or frozen, whole
½ cup sugar
1 tsp. ground cinnamon

Preheat oven to 350 degrees.

Combine all *Filling* ingredients in a bowl. Butter the bottom and sides of a 2-quart oblong (2-inch deep) glass oven-proof baking dish. Spread *Filling* on bottom of dish. Finish with *Topping*.

Bake at 350 degrees for 1 hour.

Cool on wire rack for 15 minutes. Serve warm or cold.

Makes 8 servings.

The aroma of this crisp cooking in the oven evokes, for me, childhood dreams of Granny. I hope the magic works for you.

In the early 1950's my parents left their family and friends in Pico, one of the Azorean islands, and immigrated to the United States. A couple of years later I was born. Throughout my childhood I wondered about my grandparents . . .

What were they like? Would they tell me bedtime stories? Would they play with me? Would Avó *bake goodies for me?*

The years have passed. My grandparents died without my ever meeting them. But my childhood dreams of Granny live in Granny's Cranberry Crisp.

Zesty Rhubarb Crisp

Topping

¾ cup (1 ½ sticks) chilled butter or margarine, cut into small pieces
1 cup all-purpose flour
1 Tbsp. cornstarch
½ cup packed dark-brown sugar
½ tsp. ground cinnamon
¼ tsp. ground nutmeg
⅛ tsp. salt
½ cup rolled oats

Filling

6 cups rhubarb (2 lbs or 6 large stalks) cut into ½ inch pieces
juice from one orange
1 tsp. grated orange zest
1 tsp. vanilla extract
½ cup sugar
2 Tbsp. cornstarch

Preheat oven to 375 degrees.

Use a few pieces of the cut-up butter to grease a 9- x 13-inch glass oven-proof baking dish.

Topping

In a bowl whisk flour, cornstarch, brown sugar, cinnamon, nutmeg, salt, and oats. Using a pastry cutter cut in remaining butter until mixture forms crumbs. Set aside.

Filling

Place rhubarb, orange juice, orange zest, and vanilla in a large bowl. Stir until well combined.

In a small bowl combine sugar and cornstarch. Sprinkle over the rhubarb mixture to coat all of the rhubarb.

Spread *Filling* evenly into a buttered 9- x 13-inch glass oven-proof baking dish. Cover with *Topping*.

Bake at 375 degrees for 25 to 30 minutes or until rhubarb is soft when toothpick is inserted into the center.

Transfer to cooling wire rack. Cool slightly. Serve warm.

Makes 8 to 10 servings.

Put some zest into your life. I put some into mine when it's rhubarb season. I have rhubarb growing outside my kitchen in an old whiskey barrel. When there's enough rhubarb to make this crisp, I pick it immediately, never allowing the rhubarb to age.

Whiskey is great when aged, but not rhubarb, which is best when fresh and tender.

Cranberry Christmas Cookies

4 Tbsp. (½ stick) butter or margarine, softened
1 cup dark-brown sugar
½ cup applesauce
1 large egg
2 Tbsp. molasses
2 Tbsp. water
1 tsp. vanilla extract
2 cups rolled oats
1 cup all-purpose flour
½ tsp. ground cinnamon
½ tsp. ground nutmeg
½ tsp. baking powder
¼ tsp. salt
1 cup whole cranberries, fresh or frozen
½ cup chopped walnuts

With electric mixer blend in a *large bowl* butter and brown sugar.

In a separate bowl stir applesauce, egg, molasses, water, and vanilla until combined. Add this mixture to the *large bowl* containing the blended butter and brown sugar. Beat for *two minutes*. Mixture will look "curdled."

In another bowl, using a spoon, mix rolled oats, flour, cinnamon, nutmeg, baking powder, and salt. With electric mixer on low speed, gently blend these dry ingredients into the *large bowl* containing the "curdled" mixture.

Finally, fold in cranberries and walnuts.

Cover bowl with plastic wrap and chill dough for 30 minutes.

Preheat oven to 375 degrees.

Coat cookie sheet with non-stick cooking spray.

Using an ice-cream scoop, form into perfect balls and place on cookie sheet. Bake at 375 degrees for 20 minutes, or until cookies are *golden brown*. Cool completely before enjoying.

Makes a baker's dozen. Perfect for gift-giving. You get to sample one, to make sure it's perfect, and bring the remaining dozen to your next party.

Crispy Oatmeal Raisin Cookies

1 cup rolled oats
1 cup all-purpose flour
½ cup sugar
½ tsp. baking powder
½ tsp. baking soda
½ tsp. ground cinnamon
1 large egg
⅓ cup light corn syrup
1 tsp. vanilla extract
½ cup raisins

Preheat oven to 375 degrees.

Coat cookie sheet with non-stick cooking spray.

In a *large bowl* combine oats, flour, sugar, baking powder, baking soda, and cinnamon. Mix with a spoon.

In a small bowl beat egg with a fork. Add to dry ingredients along with corn syrup and vanilla. Mix well with a spoon. Add raisins. Batter will be stiff and sticky.

With a spoon form 1-inch diameter round balls. Place on cookie sheet leaving enough room for cookies to expand. (They will triple in size.)

Bake at 375 degrees for 10 to 12 minutes, or until cookies are golden brown.

Cool cookies on wire rack.

Makes 2 dozen cookies.

"Snack."

Do you often hear someone in your home hinting for a snack? If so, these cookies will satisfy that craving for "a little something." At least, they do for my husband.

MARY'S MOCK CHERRY PIE

- 4 cups cranberries, fresh or frozen, cut in half
- 1 cup raisins
- 1 cup sugar
- 1½ Tbsp. all-purpose flour
- ⅛ tsp. salt
- ½ cup water
- 1 tsp. vanilla extract
- 1 Tbsp. butter or margarine
- ✤ pastry for a double-crust pie

Note: If you prefer sweet to tart, increase the amount of sugar from 1 cup to 1¼ cups.

In a medium size pan combine the cranberries, raisins, sugar, flour, and salt. Stir in water. Cook and stir over medium-high heat until cranberries are soft, raisins are plump, and mixture begins to boil. Continue to boil and stir for 2 minutes until mixture has the consistency of jam. Remove from heat. Stir in vanilla and butter.

Let stand to cool about 30 minutes.

Transfer to a pastry-lined 9-inch glass pie plate. Place top crust over fruit filling. Seal and flute the edge. Cut slits in the top crust.

Bake at 350 degrees for 40 minutes or until the top is golden.

Cool on a wire rack.

Makes 8 servings.

This recipe represents my mother-in-law, Mary's, many years of baking. A friend gave her a similar recipe when she was a young bride of nineteen. It was 1930, the Depression, and Mary's husband was a laborer on the cranberry bogs. Times were hard. Since cranberries were readily available, Mary often made "mock cherry" pie for her family. Over the years she modified the recipe but never recorded the exact measures. A couple of years ago I asked her for the recipe: "... a heaping spoon of flour ...a pat of butter ..."

SIMPLE PINEAPPLE PIE

- 1 can (20 oz.) *crushed* pineapple, packed in its own juice, *drained*
- 1 can (8 oz.) *crushed* pineapple, packed in its own juice
- ¾ cup sugar
- 3 large eggs, beaten with a fork
- 3 Tbsp. all-purpose flour
- ♣ pastry for a double-crust pie

Mix all ingredients (except for crust, of course) with a spoon. Include the juice *only from the 8 oz. can.*

Transfer to a pastry-lined 9-inch glass pie plate. Place top crust over filling. Seal and flute the edge. Cut slits in the top crust.

Bake at 375 degrees for 35 minutes.

Cool on a wire rack.

Makes 8 servings.

This recipe was given to my mother-in-law, Mary, in the 1930's. As a young bride and mother, she frequently made this pie, in addition to other pies, such as Mary's Mock Cherry Pie (included in this section).

Recently while reminiscing, Mary mentioned that she hadn't baked her pineapple pie in years.

I asked: "Do you still have the recipe?"

She replied: "Yes, but I'll have to look for it."

Having found the recipe, Mary baked her pineapple pie one Sunday and brought it over for dinner, along with the recipe.

... Delicious, and so easy to make.

I hope you enjoy it, too.

WILD BLUEBERRY PIE

- 4 cups wild blueberries
- 3/4 cup sugar
- 2 Tbsp. all-purpose flour
- 1/4 tsp. ground nutmeg
- 1/4 tsp. ground cinnamon
- 1 Tbsp. butter or margarine
- ✤ pastry for a double-crust pie

Pastry line a 9-inch glass pie plate.

Mix sugar and flour.

Spread one-fourth of mixture on the lined pie plate. Fill with blueberries.

Sprinkle remaining mixture over blueberries. Sprinkle nutmeg and cinnamon. Dot with butter.

Place top crust over fruit filling. Seal and flute the edge. Cut slits in the top crust.

Bake at 425 degrees for 30 to 35 minutes or until the top is golden. Juice from the berries will start to bubble over the crust.

Cool on wire rack.

Makes 8 servings.

Note: If *not* using *wild* blueberries, increase the amount of sugar from 3/4 cup to 1 cup.

Wild blueberries! A few years ago was the first time I tasted wild blueberries—thanks to our thoughtful neighbors.

"You're right! They're so sweet and juicy!" I said to my husband, who grew up in Cranberry Country, working on the cranberry bogs and picking wild blueberries to eat.

Will there be wild blueberries this year at our neighbors' special spot? With all the development—"progress" according to some—treasures like wild blueberries may become only memories for those of us who have experienced "the wild one."

Richard's Italian Rice Pie

Rice Mixture

- 2 cups cooked white rice
- 10 eggs
- 6 cups milk
- 1½ cups sugar
- 1 cup citron, chopped fine
- 3 tsp. vanilla extract
- 3 tsp. orange extract
- 1 tsp. salt

Beat eggs well. Add milk, sugar, citron, vanilla and orange extract, and salt. Stir well. Add cooked rice. Stir again. Set aside.

Pastry Dough

- 3 cups all-purpose flour
- 5 Tbsp. sugar
- 1 tsp. salt
- 3 eggs, beaten
- ½ cup (1 stick) butter, melted

Mix flour, sugar, and salt to form a dough.

Form a well in the center. Add beaten eggs and melted butter. Knead into dough.

Roll dough. Cover bottom of three 9-inch glass pie plates. Fill each pie plate with *Rice Mixture*. Cover each pie with several fine strips of dough, horizontally and vertically, resulting in a lattice look.

Bake at 350 degrees for about one hour, until the custard sets.

Makes 3 pies

Every year, friends and family of Richard Marrocco look forward to his traditional Italian Easter brunch. Of all the many gourmet items Richard includes in his brunch, my husband and I especially enjoy his Italian Rice Pie, an incredibly delicious combination of custard and rice pudding—so delicious that we have seconds and, I confess, sometimes thirds.

Richard tells us that this recipe has been handed down in his family through four generations. It was his mother, Lena, who taught Richard how to make the Rice Pies. Over the years he has made subtle changes to suit his taste.

Needless to say, this Easter tradition can be enjoyed year 'round.

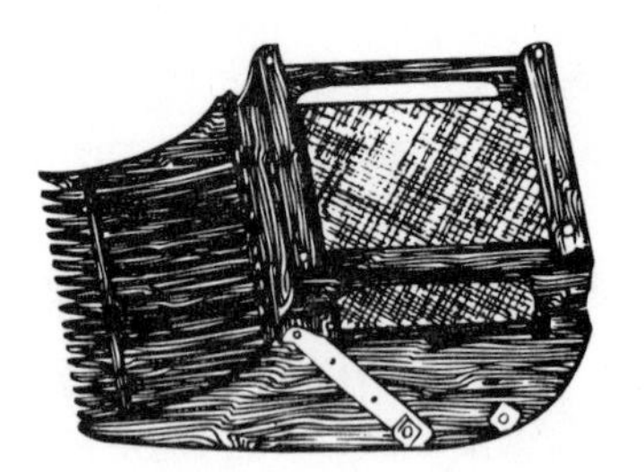

• NOTES •

• NOTES •

• NOTES •

• NOTES •